FINDING THE SPEED OF LIGHT

To my brother-in-law, Robert
Ackerman, who has encouraged me
for years and helped turn these
pages into a better story.

—M.W.

For Kaia-roo. Keep imagining, my
little vampire. Love, Aunt Bee

—R.E.

FINDING THE SPEED OF LIGHT

The 1676 Discovery That Dazzled the World

Written by
Mark Weston

Illustrated by
Rebecca Evans

Ole Romer was born in 1644 and grew up in Aarhus, Denmark. (In Danish, his name is pronounced OOH-lih ROY-mar.) Ole's father, Christen, was a sea captain. He owned a large sailboat and hauled cargo from one town to another. He taught Ole how to steer the boat at night by watching the stars. He also encouraged Ole to work hard, saying, "God does not listen to the prayers of lazy people."

PAY ATTENTION OLE! ORION'S SWORD POINTS SOUTH, AND THE BOWL OF THE BIG DIPPER POINTS NORTH.

When the town's church bells rang, Ole got up to go to school. Denmark is a cold country, so his mother, Anna, put hay in his boots for extra warmth. Ole was in a hurry most mornings because if he arrived at school even one minute late, his teacher slapped him.

Ole didn't dare talk during class, or make a funny face. But Ole's teacher liked storms. If it thundered, he stopped teaching and said, "When God speaks, I listen!"

Ole liked thunderstorms too, though some of them lasted an hour. This was a long time to sit still, but Ole loved the flashes of lightning and rumbles of thunder. He noticed that the lightning always came first, then the thunder. The light from the clouds arrived faster than the sound. "How much faster?" Ole wondered. "How fast does light travel?"

When Ole was eighteen, he went to college in Copenhagen, Denmark's capital. He learned French and took courses in astronomy, the study of the stars and planets. He was such a good student that a professor asked him to help work on some star charts that an older astronomer had made. They were the most accurate star charts in the world, but they were hard to read and needed to be redrawn. Sometimes the professor's little daughter came by to watch Ole work.

CAN ANYONE OTHER THAN OLE ANSWER THIS QUESTION?

SIGH OLE IS SO DREAMY. I THINK I HAVE STARS IN MY EYES . . .

WHAT IS IT? IT'S HUGE!

IT'S FINALLY FINISHED!

At night, Ole hurried up to his roof to look through his telescope. He built its tubes, handles, and magnifying glasses himself with money given to him by the king. When he finished, the telescope was ten feet long. On cloudy nights, all he could see was the Cathedral of Notre Dame a mile away. But when the sky was clear, Ole could see thousands of stars.

Ole particularly liked to watch Jupiter and its moons, but he was not the first person to view the faraway planet. An Italian scientist, Galileo, had also looked at Jupiter. With his own small telescope, Galileo was amazed to see that Jupiter has four bright moons circling around it.

This was an important discovery. Until then, nearly everyone had thought that the planets, including Mars, Venus, and Saturn, moved around Earth rather than the sun. But when people saw the four moons circling Jupiter, they had to admit that Earth was not the center of everything. Over time, more and more people could see that what Galileo wrote was true: the planets orbit the giant sun, not our much smaller Earth.

By the time Ole grew up, astronomers had studied the planets' orbits for years. But they knew only the shapes of these orbits, not their sizes. Astronomers did not yet know the distances between Earth, the sun, and the planets. Soon after Ole arrived in Paris, however, he and his fellow scientists made an exciting discovery. They finally learned how far away the planets are from the sun and Earth.

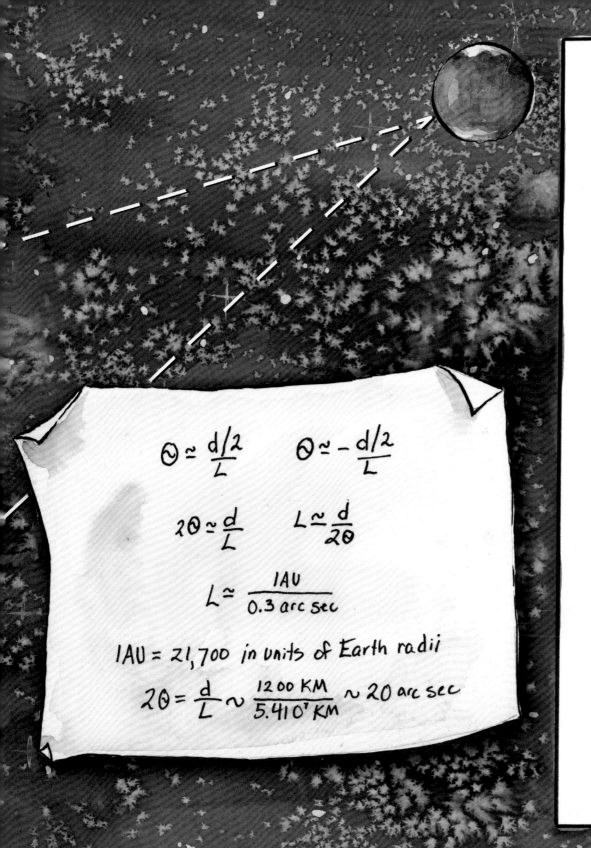

$$\Theta \simeq \frac{d/2}{L} \qquad \Theta \simeq -\frac{d/2}{L}$$

$$2\Theta \simeq \frac{d}{L} \qquad L \simeq \frac{d}{2\Theta}$$

$$L \simeq \frac{1AU}{0.3 \text{ arc sec}}$$

$$1AU = 21,700 \text{ in units of Earth radii}$$

$$2\Theta = \frac{d}{L} \sim \frac{1200 \text{ KM}}{5.4 10^7 \text{ KM}} \sim 20 \text{ arc sec}$$

Astronomers in Paris worked with astronomers in Cayenne, a French city in South America 5,000 miles away, to figure out the distance to Mars.

They saw that Mars's place in Paris's night sky is slightly different from Mars's place in South America's night sky. Using this tiny difference in positions, the scientists pretended that there was a giant triangle linking Paris, South America, and Mars.

The astronomers knew that the length of the shortest side of this triangle (the side from Paris to Cayenne) was 5,000 miles. And because they also knew the exact angles upward from each city to Mars, they could finally measure the longer sides of this triangle—the distance between Earth and Mars.

Ole and his friends saw that the distance between Earth and Mars changes a lot as the two planets circle the sun. When Earth and Mars are on the same side of the sun, Mars is about 35 million miles away. How far is this? Imagine that you could drive a car through space at 65 miles per hour without stopping. It would take you 62 years to reach Mars. But this is only when Mars is on the same side of the sun as Earth. When Earth and Mars are on opposite sides of the sun, Mars is seven times farther away!

Because Earth goes around the sun almost twice as quickly as Mars does, circling once a year when Mars takes nearly two years, astronomers had long ago used a kind of math called geometry to figure out that Earth is about two-thirds as far away from the sun as Mars is. So now that Ole and his friends knew how far Mars is from Earth, they could also figure out how far the sun is from Earth.

Ole thought the sun was 86 million miles from Earth, although today we know that it's really 93 million miles away. How far is this? Imagine again that you could drive a car through space at 65 miles per hour without stopping. It would take you 163 years—two lifetimes!—to drive from Earth to the sun.

I THOUGHT ASTRONOMERS ARE JUST SUPPOSED TO LOOK AT STARS, WHY DO WE HAVE TO DO SO MUCH MATH?

ARE WE THERE YET?

ONLY 126 YEARS TILL WE ARRIVE, SON!

Jupiter, the planet Ole most liked to watch, is five times farther away from the sun than Earth is, a distance of more than 480 million miles. If we added jet engines to our space-car, it would still take 100 years to fly to Jupiter.

Even on freezing winter nights, when Ole shivered and drank tea to keep warm, he loved to look through his telescope and see Jupiter's brown and white stripes. Jupiter, named after the supreme Roman god, is a gigantic ball of ice-cold gas, the biggest planet in our solar system. More than 1,300 Earths could fit into Jupiter.

BLUE SNOW? WHOEVER HEARD OF BLUE SNOW?!

OF COURSE, BLUE SNOW! THAT MEANS IT MUST BE REALLY, REALLY COLD, RIGHT?

BLAM!

PFFFT!

Ole watched Io every night (unless the skies over Paris were cloudy). During part of Io's orbit, the moon disappears behind one side of Jupiter, then appears again on the other side two hours later. Ole liked trying to predict the exact minute when Io would pop out from behind Jupiter. "Yes!" he said to himself when he called the time correctly.

Ole cleaned every part of his telescope. He checked every part of his clock. He read his notes all over again. He looked at observations of Io that an older astronomer, Giovanni Cassini, had made eight years before.

But Io's movements remained a mystery.

Ole kept watching Io, but the time that Io seemed to take to go around Jupiter changed by just a few seconds from one orbit to the next, and Ole's clock was not accurate enough to measure such small differences. Ole hoped instead that if he observed Io carefully, for a long time, he might see a pattern.

He did.

Ole saw that every month, for six straight months, Io seemed to slow down and take about 3½ more minutes to orbit Jupiter than it had taken the month before. So after six months Io seemed to take 22 minutes longer to circle Jupiter than it had taken half a year earlier.

Then Io appeared to speed up again. Every month, for six months in a row, Io seemed to take about 3½ fewer minutes to orbit Jupiter than it had taken the previous month. After six months, the time that Io took to circle Jupiter appeared to be 22 minutes shorter than it was before.

"This can't be real," Ole said, shaking his head. "It has to be an illusion. But why does Io seem to speed up and slow down?"

WAIT! I'VE GOT IT! LIGHT ISN'T INSTANT. IT MOVES, AND I CAN MEASURE ITS SPEED!

I HAVE IT! IT TAKES LIGHT 22 MINUTES LONGER TO REACH EARTH WHEN WE'RE ON THE OPPOSITE SIDE OF THE SUN FROM JUPITER!

EARTH CLOSER TO JUPITER

NEPTUNE

VENUS

URANUS

JUPITER

And then it hit him. Io appeared to speed up and slow down, Ole realized, not because of anything that was happening near Jupiter or near Io every six months, but because the distance between Earth and Jupiter shrinks when both planets are on the same side of the sun, and grows when the two planets are on opposite sides of the sun.

When Io seemed to take 22 minutes more to orbit Jupiter than it had taken just six months before, it was because Earth was on the opposite side of the sun from Jupiter and Io, and light needed more time to cross this longer distance.

"So light is not instant," Ole said. "It moves, and I can measure its speed."

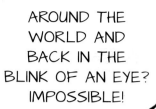

ROYAL ACADEMY
OF SCIENCE
Today's speaker:
Ole Romer discovers
light has a speed!

AROUND THE WORLD AND BACK IN THE BLINK OF AN EYE? IMPOSSIBLE!

I DON'T BELIEVE IT!

When the Royal Academy of Science announced Ole's discovery in 1676, many people did not believe that light could travel so quickly. "Around the world and back in the blink of an eye? Impossible!" they said.

But Ole's math and science were hard to argue with, and before long nearly everyone accepted his conclusion: Light has a speed, and that speed is incredibly fast.

IN CONCLUSION, LIGHT TRAVELS AT 130,000 MILES PER SECOND!

IMPOSSIBLE!

RIDICULOUS!

I STILL THINK I SHOULD BE THE CENTER OF THE UNIVERSE!

NEVER!

WAIT, HIS MATH MAKES SENSE!

In fact, light is even more rapid than Ole thought. Only twenty-five years after his discovery, astronomers with better telescopes learned that the sun is 93 million miles away from Earth, a little farther than Ole had thought. And because they could see Io's movements more clearly than Ole could, they learned that Io's orbit around Jupiter actually seems to speed up and slow down by 16½ minutes every six months, not 22 minutes as Ole had believed.

Light takes only 16½ minutes to travel the 186 million miles that stretch from one side of Earth's orbit around the sun to the other. Sixteen-and-a-half minutes is almost 1,000 seconds, so light travels more than 186,000 miles (299,000 kilometers) per second!

OLE ROMER WAS RIGHT, LIGHT DOES HAVE SPEED, BUT IT'S EVEN FASTER THAN HE THOUGHT!

EVEN THOUGH HIS TELESCOPE WAS TEN FEET LONG, IT WAS STILL PRETTY SIMPLE. IT'S AMAZING THAT HE FIGURED IT OUT AT ALL!

How fast is this? Light is more than 100,000 times faster than a bullet, and a bullet is 100,000 times faster than a snail. So compared to light, even the fastest bullet is as slow as a snail!

AFTERWORD

Today, thanks to the work of Albert Einstein (1879–1955) we know that distance and time are "relative" and can change as one's speed increases, but the speed of light in a vacuum is always the same. Whether light comes from a star or a light bulb, a space ship or a firefly, its speed is 186,282 miles (299,792 kilometers) per second, never any faster. In fact, *nothing* travels faster than light, although electricity, x-rays, and radio waves move equally as fast. (Light slows down by a quarter when advancing through water instead of a vacuum, by a third when passing through glass, and by more than half when moving through a diamond—just as electricity slows down when traveling through a wire.)

Even at such an incredible speed, light takes a long time to travel from one star to another through the vast expanse of space. The nearest star to Earth, Proxima Centauri, is 25 trillion (that's 25 million million!) miles away, and its light takes over four years to reach us. Our Milky Way galaxy—which includes all the stars we can see in the night sky with our naked eyes—is so vast that light from its center travels 30,000 years before it finally reaches our solar system in an arm of the spiral-shaped galaxy. Other galaxies are so distant that their light takes millions or even billions of years to reach us.

In movies and on television, starships travel at "warp speed," thousands of times faster than the speed of light. This is fun to watch, but it isn't real. At least as far as twenty-first century science is concerned, the speed of light is the universe's firm, unbreakable speed limit.

To answer the question that young Ole asked himself during thunderstorms, light travels almost a million times faster than sound. This is why a bolt of lightning one mile away is visible immediately, but the thunder it produces will take five seconds to reach you. That Ole Romer could figure this out in 1676, when there were no cameras or computers to help him, is almost as amazing as the speed of light itself.

OLE'S LATER YEARS

Ole Romer returned home to Denmark to become his country's chief astronomer. He built bigger and better telescopes and smaller and more accurate clocks. He helped improve Denmark's ships and windmills and helped a German glassblower, Gabriel Daniel Fahrenheit, invent a better thermometer.

When Ole was thirty-seven he married Anne Bartholin, the daughter of the professor he had drawn star charts for. In 1705 he became Copenhagen's chief of police. He cut crime by installing oil lamps to light the city's streets at night. Five years later, in 1710, Ole Romer died at sixty-five. He was one of the greatest astronomers of all time, and the first person to measure the speed of light, the fastest thing in the universe.

TIMELINE

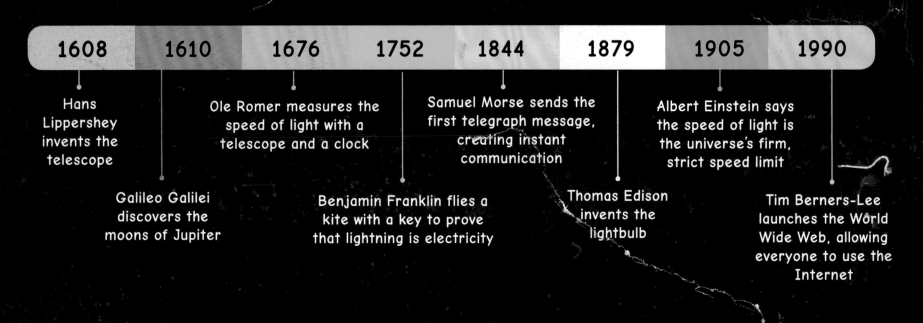

| 1608 | 1610 | 1676 | 1752 | 1844 | 1879 | 1905 | 1990 |

Hans Lippershey invents the telescope

Galileo Galilei discovers the moons of Jupiter

Ole Romer measures the speed of light with a telescope and a clock

Benjamin Franklin flies a kite with a key to prove that lightning is electricity

Samuel Morse sends the first telegraph message, creating instant communication

Thomas Edison invents the lightbulb

Albert Einstein says the speed of light is the universe's firm, strict speed limit

Tim Berners-Lee launches the World Wide Web, allowing everyone to use the Internet

MARK WESTON graduated from Brown University with a B.A. in History and earned a law degree from the University of Texas. He has been a lawyer for ABC Television and a journalist for ABC News. In 1991, Mark won enough money on TV's *Jeopardy!* to start a company making geographical jigsaw puzzles for children, which he sold after three years. His history books include *Giants of Japan* and *Prophets and Princes: Saudi Arabia from Muhammad to the Present*. His children's books include *The Story of Soichiro Honda*. Mark lives in Sarasota, Florida, with his wife, painter Linda Richichi. His website is *markwestonauthor.com*.

REBECCA EVANS worked for nine years as an artist and designer before returning to her first love: children's book illustration and writing. She has illustrated fourteen picture books (including *Masterpiece Robot*, Tilbury House, 2018) and nine middle-grade readers and has authored one picture book. She lives in Maryland with her husband and four young children, shares her love of literature and art regularly at schools, teaches art at a performing and visual arts school, is a regional advisor in the SCBWI, and works from her home studio whenever time permits. Rebecca's boundless imagination enjoys free rein at *www.rebeccaevans.net*.

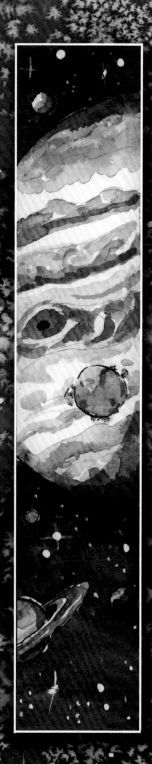

I would like to thank Mogens Lehmann for sharing his novel about Ole Romer, *The Hesitation of Light*, before it was published; Dr. Jacob Howland for directing me to the autobiography of Danish philosopher Soren Kierkegaard, which proved invaluable when writing about Romer's childhood (about which we know almost nothing); and my Danish friend Jakob Schroeder for his years of encouragement.

—M.W.

Tilbury House Publishers
12 Starr Street
Thomaston, Maine 04861
800-582-1899 • www.tilburyhouse.com

Hardcover ISBN 978-0-88448-545-2
ebook ISBN 978-0-88448-547-6

First hardcover printing March 2019

15 16 17 18 19 20 XXX 10 9 8 7 6 5 4 3 2 1

Library of Congress Control Number: 2018960099

Cover and text designed by Frame25 Productions
Printed in China through Four Colour Print Group, Louisville, KY